Broward Marine Fire
Ft. Lauderdale, Florida

Investigated by: Dennis C. Duckett

Written by: Sheila-Faith Barry

This is Report 101 of the Major Fires Investigation Project conducted by Varley-Campbell and Associates, Inc. under contract EMW-94-C-4423 to the United States Fire Administration, Federal Emergency Management Agency.

 FEMA

Department of Homeland Security
United States Fire Administration
National Fire Data Center

U.S. Fire Administration Fire Investigations Program

The U.S. Fire Administration develops reports on selected major fires throughout the country. The fires usually involve multiple deaths or a large loss of property. But the primary criterion for deciding to do a report is whether it will result in significant "lessons learned." In some cases these lessons bring to light new knowledge about fire--the effect of building construction or contents, human behavior in fire, etc. In other cases, the lessons are not new but are serious enough to highlight once again, with yet another fire tragedy report. In some cases, special reports are developed to discuss events, drills, or new technologies which are of interest to the fire service.

The reports are sent to fire magazines and are distributed at National and Regional fire meetings. The International Association of Fire Chiefs assists the USFA in disseminating the findings throughout the fire service. On a continuing basis the reports are available on request from the USFA; announcements of their availability are published widely in fire journals and newsletters.

This body of work provides detailed information on the nature of the fire problem for policymakers who must decide on allocations of resources between fire and other pressing problems, and within the fire service to improve codes and code enforcement, training, public fire education, building technology, and other related areas.

The Fire Administration, which has no regulatory authority, sends an experienced fire investigator into a community after a major incident only after having conferred with the local fire authorities to insure that the assistance and presence of the USFA would be supportive and would in no way interfere with any review of the incident they are themselves conducting. The intent is not to arrive during the event or even immediately after, but rather after the dust settles, so that a complete and objective review of all the important aspects of the incident can be made. Local authorities review the USFA's report while it is in draft. The USFA investigator or team is available to local authorities should they wish to request technical assistance for their own investigation.

This report and its recommendations were developed by USFA staff and by Varley-Campbell & Associates, Inc., Miami and Chicago, its staff and consultants who are under contract to assist the Fire Administration in carrying out the Fire Reports Program.

The U.S. Fire Administration greatly appreciates the cooperation and information received from officials of the Fort Lauderdale Fire-Rescue and Building Department.

For additional copies of this report write to the U.S. Fire Administration, 16825 South Seton Avenue, Emmitsburg, Maryland 21727. The report is available on the Administration's Web site at http://www.usfa.dhs.gov/

U.S. Fire Administration

Mission Statement

As an entity of the Department of Homeland Security, the mission of the USFA is to reduce life and economic losses due to fire and related emergencies, through leadership, advocacy, coordination, and support. We serve the Nation independently, in coordination with other Federal agencies, and in partnership with fire protection and emergency service communities. With a commitment to excellence, we provide public education, training, technology, and data initiatives.

TABLE OF CONTENTS

Broward Marine Fire
Fort Lauderdale, Florida
September 5, 1996

Local Contacts: Stephen R. McInerny II
Division Chief

Tony Precanico
Lieutenant, Fire Investigator

Bruce Strandhagen
Fire Inspector

Fort Lauderdale Fire-Rescue and Building Department
300 Northwest 1st Avenue
Fort Lauderdale, Florida 33301-1098

James T. Pott, Police Detective

Fort Lauderdale Police Department
1300 West Broward Boulevard
Fort Lauderdale, Florida 33312

OVERVIEW

An early morning fire destroyed the Broward Marine boat manufacturing facility in Fort Lauderdale, Florida, and caused extensive damage to several yachts under construction at the facility. The firestorm spread large embers throughout the area, causing spot fires throughout the property and damaging several boats moored near the site. The resulting smoke column was visible from as far as seven miles away.

Situated on the New River, the Broward Marine facility housed boat production and storage buildings dating back to the 1930's and marina areas. The structures involved in the fire included a manufacturing building, a boat assembly building, and an office building. (See Figure 1) The facility had previously been cited by authorities for a number of code violations. Damage was estimated at 15 million dollars.

The facility's security guard placed a call to 9-1-1 shortly after midnight, reporting building in flames. The first arriving fire company, Engine 3, found the roof of the manufacturing building (Building A) had collapsed before their arrival on scene. The first responding company officer called for a second alarm while enroute to the fire. Heavy black smoke was visible from fire station no. 3 located approximately 22 blocks away. A third alarm was called immediately upon arrival. The fire eventually required six alarms and mutual aid support from the Coast Guard, Port Everglades and five neighboring fire departments.

Firefighting efforts were hampered by the difficulty of accessing the site. West side access was restricted by a canal; the New River was located on the north side of the facility. The main access road was a narrow residential street partially blocked by parked private vehicles. Several covered boat slips abutted the property on both the north and west boundaries of the site.

The facility had no fire detection or alarm system and no automatic fire suppression system. Water supply was from one yard hydrant and two hydrants off property, which were all supplied from the same main. Drafting was not possible on the north side of the fire due to the shallow waters and was impractical on the west side, as the fire building's west wall was built up to the canal. Fort Lauderdale's fire boat (FB 49) was pressed into service to feed supply and handlines on the north side of the facility.

The building of origin (Building A) contained two yachts under construction and a large inventory of materials used in the manufacture of the yachts, including teak planks, fiberglass fabricating supplies and flammable liquids stored in barrels. Building A also housed a paint booth and a mezzanine area containing various storage.

Exposures were an immediate concern and included an assembly building (Building B) and a two story office building (Building C), both abutting the building of fire origin. Building B contained two 100-foot yachts in the final stages of construction. Also located on the property were a 1500-amp electric service distribution system, a dust collector associated with a woodworking shop, metalworking and paint shops. Several older wooden structures situated around the property housed storage and office areas.

As the firestorm developed, softball sized embers spread from the fire and ignited numerous spot fires around the property. Several boats in a neighboring marina on the east boundary of the property were damaged by firebrands. Although the westerly wind helped to protect the boats on the canal and to a lesser extent those in the north slips, all the boats were at risk until the fire was extinguished.

KEY ISSUES

Issues	Comments
The Facility	Constructed prior to World War II, the facility's buildings were of unprotected ordinary construction. Improper storage of flammables, poor housekeeping practices in manufacturing areas, and non-compliance with fire codes created significant risk of structural collapse and exaggerated firefighter hazards.
Delayed Alarm	There was no fire detection and alarm system and the fire went unreported until it had become well involved.
Fire Origin	The fire was determined to have been caused by an electrical event stemming from a failure in an electric panel box in the manufacturing building. Fire department inspectors had previously cited electrical problems.

Issues	Comments
Tactical Considerations	Because of the time of the incident, fire ground officers had no plant personnel to consider. Restricted access and spread of the fire due to significant radiant heat and a fully developed firestorm was the most pressing problem. Officers directed immediate deployment of heavy streams and established perimeters for firefighters. Firefighters quickly set up heavy streams, limiting spread to immediate exposures and spot fires which were controlled by secondary handlines.
Environmental Concerns	Concern for run-off of various flammable liquids and hazardous chemicals was cause for deployment of floating booms to lessen contamination of the riverway. Some flammable chemical fires were contained and allowed to burn down, since tactical operations had limited their contribution to the overall fire.
Water Supply	Officers addressed water issues early in the fire as draft operations were considered and subsequently ruled out. A pressure increase of the distribution system was requested; the water department responded promptly but two main breaks resulted from the increase of line pressure. Fireboat operations eased the water supply concerns.
Mutual Aid	Established mutual aid procedures with various emergency response providers assured prompt response to needs. Department procedures were in place to assure adequate staffing despite the extraordinary demands of the fire.

THE FIRE DEPARTMENT

Fort Lauderdale Fire-Rescue and Building Department protects a city of 33 square miles with a population slightly in excess of 150,000. The department has 290 uniformed firefighters and 12 stations. They run paramedic/engineers on fire apparatus, with advanced life support and basic life support units in the emergency medical services division.

The department runs a total of eleven engine companies, three ladder companies, one air support and light truck, one airport crash truck located at Fort Lauderdale Executive Airport, four medical rescue units, one EMS Coordinator, three district battalion chiefs, and one division chief. These companies and vehicles or positions are staffed on a daily basis.

Eight of these engines are ALS equipped and ride with at least one paramedic onboard. The medical rescue units are staffed with at least two paramedics. The Department provides ALS and BLS medical response and transport services. All twelve stations operate within the Operations Division, providing fire suppression, protection and emergency medical care.

Fort Lauderdale Fire-Rescue has formal mutual aid agreements with all cities located within Broward County. They routinely respond to neighboring cities for mutual aid. Each fire-rescue agency has a copy of the mutual aid plan. Mutual aid requests are handled by the Broward County Fire-Rescue Department Communications Center.

The department routinely drills with other departments countywide and Regionally pertaining to hazardous materials and technical rescue. The department operates both hazardous material and technical rescue teams; the technical rescue team specifically handles underwater dive rescue, confined space and elevated victim rescue.

The fire department inspection division is responsible for all permitting, review and inspection functions within the city. The Building Department operates as a division of the fire department.

THE BUILDING AND ITS OPERATIONS

The Broward Marine, Inc. facility was located in the incorporated city limits of Fort Lauderdale, Florida. The company originally built P.T. boats, minesweepers and sub tenders for the U.S. military. At the time of the fire it produced aluminum-hull luxury yachts. Repair and docking services were also provided. Approximately 30 yachts, over 100 feet in length, were stored on the property in docks that ran along the west and north perimeters of the property. Most of the buildings at the site dated back to the 1930s, although some updating and minor new construction had occurred over the years. Railway lines ran through the property, with tracks laid through Building A and Building B from north to south.

The largest building on the site was the manufacturing building (Building A), which abutted the canal directly to the west. Adjoining Building A were the one story assembly building (Building B) and the two story office building (Building C). A 30 foot tall dust collector was located directly to the east, in close proximity to the main electrical supply for the complex and a building housing restrooms and the electric winch for the boat launch. (See Figure 1) The one story, 90 by 140 ft woodworking shop and the 70 by 80 ft two story machine shop were situated to the east of the dust collector. The woodworking and machine shops incurred minimal damage as a result of the fire.

Several smaller wooden buildings were located around the site, including a sales office, storage buildings and paint and repair shops. A small wooden building used for the storage of flammable liquids was located on the far east end of the site, removed from the fire area.

Covered boat slips, a boat launch and a boatlift were located towards the north end of the property. Located around the property were dumpsters, storage sheds and fuel pumps, in addition to aluminum boat hulls in various stages of construction.

The main entrance to the site was located on Southwest 20th Street via a driveway that passed by the security shack. The driveway continued north through the property to the main entrance of the office building. A parking lot was situated to the east of the main entrance within the property's fenced perimeter.

The plant employed approximately 200 people who worked during the day. There was a security guard on duty during off hours. The plant was not in operation at the time of the fire; the one security guard on duty was posted at the guard shack.

CONSTRUCTION OF COMPLEX BUILDINGS

Manufacturing Building--Building A

The building of origin was a two story, wood and metal pole style structure used for boat manufacturing. Measuring 300 by 150 feet, the north and south ends were open, and railway tracks ran through the building through the open ends. Yachts were moved on the railway tracks within the building and outside to storage located on the north side of the building. An open mezzanine, approximately 15 feet above grade, ran on both sides of the building from north to south.

The roof of the manufacturing building was of wood truss construction; several termite damaged trusses had previously been cut away from the center of the building span and steel I-beams had been placed across the open span. Steel straps had been installed around the remaining I-beams and the remaining trusses. The I-beams were supported by the original creosoted pole columns.

A paint booth was situated on the west side of the building, under the mezzanine. An electrical circuit box was located in the mezzanine area on the far west end.

At the time of the fire, two 100-foot yachts under construction were located in the two west end bays of Building A.

Assembly Building--Building B

Adjoining the manufacturing building to the east was a two story in height, wood and metal building used for the final assembly and rigging of yachts under construction. This assembly building measured 100 x 140 feet. Two yachts were stored in Building B at the time of the fire. Both of these yachts were surrounded by wooden scaffolding. A finished yacht had been moved to the yard outside, just to the east of Building B.

Approximately six years prior, a wall had been constructed separating Building A from Building B and Building C to the east. The wall was intended as a two-hour fire wall; however, there were windows opening from the mezzanine level of Building A over the roof of Building B.

Office Building--Building C

A two story building (Building C) adjoined Building B to the south. This building, measuring 100 x 45 feet, was of unprotected wood construction and housed the facility's main offices.

THE MANUFACTURING OPERATION

The manufacturing process included aluminum and fiberglass casting, painting, varnishing, woodworking and metalworking activities occurring in different areas throughout the site.

Materials used in the construction and outfitting process represented various hazards. There were numerous oxy/acetylene rigs, perchloric acid, varnish, solvents, fiberglass sheeting and resin stored in various locations throughout the complex in addition to supplies maintained in Buildings A and B. Also stored in Building A were 6-7 foot tall stacks of teak, cypress and marine plywood as well as a large stock of canvas.

A small berm had been constructed around a few old marine cargo containers, which were used as storage lockers for flammable liquids. These cargo containers were located inside Building A.

There was no established procedure for cleaning of work areas. Materials and supplies were routinely left out in the work area. The facility had previously been cited by the fire department on several occasions for such poor housekeeping practices.

FIRE PROTECTION

There were no fire alarm, fire detection or automatic sprinkler systems installed in any of the buildings on the site. However, an on-site fire pump had been installed to supply hose reels in the manufacturing building.

The hydrant located at the intersection of Southwest 20th Street and Southwest 15th Avenue had a statis pressure of 72 psi and a residual pressure of 65 psi at a flow of 1300 gpm. This hydrant is supplied through approximately 150 feet of 12-inch underground pipe, which is connected to a 20-inch city water main located under Southwest 15th Avenue. This 20-inch city water main also supplied a 6-inch loop to the west, which ties back into the 20-inch water main to the south. (See Figure 3)

INSPECTIONS AND CODE COMPLIANCE

For approximately 30 years prior to the incident, unpermitted electrical work had been done on site by plant electricians; fire department inspection records cite electrical work that did not conform to codes. Tests performed by the local utility provider indicated that the plant was drawing as much as 2600 amps from the 1500 rated amp service.

Inspection records on the facility date back to 1966 and indicate several recurring problem areas, including missing or uncharged extinguishers, poor housekeeping and various problems with electrical wiring. The Inspection Division had developed a comprehensive plan for an upgrade of the entire plant at the time the fire occurred.

THE FIRE

At approximately 12:52 a.m. a security guard on duty at the facility heard a boom and then noticed flames near the rear of the facility. He placed a call to 9-1-1 to report the fire at 12:52 a.m., but he was not confident of which building was burning; he paused during the call to look up the facility's address. After placing the call, the telephone line went dead; the security guard then left the guard shack to awake the facility's owner, whose home was located across the canal to the northwest of the complex.

A second call reporting the fire came in to the 9-1-1 center at 12:54 a.m. The caller described hearing explosions and observing sparks and flames threatening homes located on the south side of the facility.

The occupant of a mobile home located across the canal at the west end of the property was awoken shortly before 1:00 a.m. by loud popping noises. Looking out his bedroom window, he saw flames coming from the southwest corner of Building A. After waking his roommate, the witness attempted to call 9-1-1, but the telephone line was dead. As he went to awaken the facility's owner, who lived just north of the trailer, he noticed a fire smoldering on the wooden docks adjoining the property.

Two witnesses on a boat moored at the marina directly north across the New River from the facility noticed flames on the west wall of the manufacturing building. They watched as the fire quickly spread towards the east.

A resident who lived in a home across Southwest 20th Street from the boatyard awoke to bright orange light flooding her bedroom. She watched as the fire quickly grew in size, reporting popping noises and flying embers.

Fire Department Dispatched

The first alarm was at 12:52 a.m. and Engine 2, Quint 2, Engine 3, Engine 8, Battalion Chief 102 and Battalion Chief 402 were dispatched. (See Appendix C for breakdown of the six alarms.) Engine 3 requested a second alarm prior to arrival. Engine 29, Engine 49, Ladder 35, the air support unit - Support 13, Battalion Chief 302, and Division Chief 2 were dispatched on the second alarm.

Upon departing quarters at Station 2 on Andrews Avenue, 37 blocks away, a column of heavy black smoke was observed in the vicinity of the marina. Battalion Chief 102 requested a third alarm transmitted at 12:59 a.m. Responding on the third alarm were Engine 35, Engine 47, and Ladder 49. Mutual aid response on the third alarm were the Port Everglades fire boat and the Coast Guard fire boat.

A police unit already on the scene reported heavy black smoke. Calls describing heavy smoke, flames and explosions continued to come into 9-1-1; dispatch updated all companies en-route of reports of explosions and large flaming embers floating over the area.

Initial Attack

On arrival at 12:58 a.m., the company officer on Engine 3 reported heavy flames extending 100 feet in the air and numerous explosions from within Building A. The roof of Building A had collapsed, and flames were spreading to Building C. Engine 3 focused initial attack on Building C, staging their apparatus to the east of the building.

Quint 2 arrived on location and was ordered to Engine 3's location to set up the aerial for deluge operations on Building B. A 5 inch supply line was laid by Engine 2 to supply Quint 2 from the hydrant located at Southwest 17th Avenue at 20th Street, on the southeast corner of the property. After supplying the line to Quint 2, the crew of Engine 2 moved the apparatus out of the immediate area. Quint 2 was set up at the southeast side of Building C with a deluge gun, a 3 inch master stream and a 1-3/4-inch handline.

Battalion Chief 102 arrived at 1:01 a.m. and assumed command from the company officer on Engine 3. Command was located at the southwest corner of the property. Quint 2 and Engine 3 were committed on the south side of the fire; it was decided that no other apparatus would move inside this perimeter. The decision was made not to evacuate the nearby homes; however, third alarm companies were assigned to check the marina to the east as numerous embers were seen falling into that area.

Upon arrival, Engine 29 laid two supply lines to the southwest of Building A and two supply lines into the yard in the vicinity of Engine 2. The crew of Engine 29 then set up on southwest side of the Building A with a deluge gun. The heavy caliber stream was established to knock down the heavy concentration of fire on the southwest side of the marina and also to protect the vessels moored on the west side of the canal.

Engine 3 set up a portable deluge gun on the east side of Building B to protect the yacht stored in the yard to the east. The crew of Engine 3 also assisted the crews of Engine 2 and Engine 8 with stretching 3 inch handlines into the same area.

While responding, Division Chief 2 observed heavy flames and numerous explosions over the facility. A firestorm was in progress and the flames were quickly spreading from Building A into Building B. Division Chief 2 called for the evacuation of all vehicular traffic from the immediate area. It is notable that the only street access to the facility was a narrow residential street. Although the property is sizable, it would have taken a few vehicles to block or severely hinder fire ground operations.

Division Chief 2 ordered command relocated from the southwest to the southeast corner of the property and requested that dispatch notify all off-duty chief officers via pager of a major multiple alarm fire and to report to the scene. Battalion Chief 402 was assigned to the position of Safety

Officer. The safety officer was assigned to survey north of Buildings A and B; north and south sectors were established.

Fire Spreads to Building B

As the fire spread, the two yachts in the open sided Building B were threatened. The firestorm was intensifying, with large embers floating into neighboring areas, and whirlwinds of flame were visible above the marina. Witnesses watching the fire from outside their homes reported high heat and feeling the "vibration" of the fire. With the threat of fire spreading from the fully involved Building A and floating firebrands, Command requested a boost in city water pressure.

Engine 3 informed Command that radiant heat was igniting new fires on the east side of Building B, specifically the dust collector. The crew of Engine 3 set up an elevated stream on Building A and requested a 5 inch supply line. Command ordered Engine 2 to stretch two 3 inch supply lines from Quint 2 to Engine 3. Quint 2 had the aerial set up on the southeast corner of Building C and was providing water supply to Engine 3's deck gun; there were two 3 inch supply lines from the yard hydrant located at the northeast corner of the property. Quint 2 was also pulling 3 inch exposure lines from the apparatus.

When Ladder 49 arrived on scene; they were returned to quarters and directed to respond with Fort Lauderdale's fireboat (FB 49), the first of three marine units to operate at the incident.

Flying brands were landing on the roofs of the facility's buildings located to the east of the fire, igniting some roofs. Brands were also igniting spot fires in dumpsters and in a row of palm trees that bordered the docks on the east perimeter. The crew of Engine 8 was directed to the dock area across the canal to survey for and extinguish spot fires.

Engine 29 reported that they were attacking a very heavy volume of fire on the southwest side of Building A but the canal located to the immediate west of the building would prevent the fire from overrunning their position. The crew directed an unmanned deck gun on the southwest corner of the building. Radiant heat did not appear to threaten the boats docked across the canal.

A yacht located in the yard on the east side of Building B was threatened as the size of the fire grew and the volume of firebrands increased. Command ordered a 3 inch supply line taken up the driveway to the immediate east side of Quint 2's position to supply more maneuverable 1-3/4-inch handlines; these lines were used to protect the yacht stored in the yard and other yachts moored at the docks north of Building B.

Fort Lauderdale's fireboat (FB 49) responded up the New River and reported to the north side of the facility to supply attack lines. The deluge gun on the boat was used to darken down the fire in Building A, which was threatening the two yachts in the adjoining Building B.

Command directed the responding Coast Guard fireboat to the north perimeter of the facility.

Second Entrance to Facility Located

The crew from Engine 2 discovered an access road on the extreme far east side of the property. This gated road, located near the staging area, had once been used to drive heavy equipment into the marina but had not been used in some time. Trees and heavy foliage had effectively hid the entrance from view. Upon entering through the gate, firefighters noted the hazard of two gasoline tanker trucks, and numerous spot fires in the area. Three dumpsters were reported burning further down the access road. A warehouse located at the far northeast corner of the property, between the fire and the New River, was threatened. Engine 49 was ordered to the area to protect this exposure.

Engine 29 reported that the elevated stream was allowing them to make good headway against the fire on the west side of Building A. Receiving verification from Engine 29 that they could hold their area, Command committed all remaining crews to the east side of the fire. The safety officer, also functioning as North sector command, called that they did not have enough water in Building B and the west side yacht was now burning. The need now was for a line between the two yachts to try to save the yacht in the east bay of Building B.

Engine 49 was trying to get to the north side location near the fireboat since the boat was now set up to supply water on the north side. Engine 49 set up with two lines to protect the yacht in the east bay of the assembly building; Engine 47's crew hand laid an attack line down the drive to the north side using 1000 feet of 3 inch hose with a highrise pack. The fireboat was ordered to shut down their deck gun and divert supply to handlines.

An emergency call went out to Q2 to shut down the deluge gun due to a burst length of hose at approximately 1:52 a.m. A fourth alarm was called at 1:53 a.m. Two engines, Engine 46 and Engine 13, responded on the fourth alarm.

Ladder 35 staged at Riverbend Marina just to the east of the fire; several spot fires were burning in the marina.

The dust collector was ignited by radiant heat early in the fire. When the fire was initially observed there was concern that the fire might endanger the electrical distribution station located to the immediate southeast or weaken the steel structure of the dust collector itself, but this was determined not to be a concern after evaluation. Some water was applied to the dust collector but it was not a focus of attack, as the fire was smoldering and contained.

Fort Lauderdale's fireboat was supplying four handlines for the attack on the yachts located inside Building B, which was still intact.

Access to the yachts in Building B was hampered due to the scaffolding erected around the yachts. Walkways laid across the scaffolding blocked the flow of water to the west end of the far most yacht, and flames were spreading.

A foam educator was set up on the wyed line supplied from Fort Lauderdale's fire boat (FB 49); firefighters used two 1-3/4-inch foam lines on the burning yacht in an attempt to stop the spread of flames.

At one hour and fifteen minutes into the incident, command reported very heavy fire with five master streams in operation and at least four handlines in operation. On the north side, Fort Lauderdale's fire boat was supplying a 3 inch line that crews were using to divide the fire east and west, and was also supplying two handlines for use on the burning yacht in the west bay of Building B.

Drafting from remote positions was ruled out because of the need to cross fences with hand laid lines and the demand on manpower that would be required as well as the risk of being trapped if the fire should jump lines.

The Coast Guard fire boat on the north perimeter supplied one 3 inch line used to extinguish spot fires to the west of Building A. The Port Everglades fire boat (FB 6) arrived on scene and directed their deck gun on the west end of Building B.

As water was applied to the yachts in Building B, there was a concern that the scaffolding would not be capable of supporting the weight of both a yacht and the firefighting water the vessel could collect. The Safety Officer directed crews away from the yacht located in the west bay to reduce the risk from a potential collapse. Facility representatives were unable to provide definitive information regarding the weight that the support jacks and scaffolding could support. It was decided to move the master stream off the area; two 1-3/4-inch handlines were then used to protect the yachts. At this time, the fire on the southwest perimeter had been knocked down, and the elevated stream was relocated to the southeast corner of Building C.

Fifth Alarm

Command called for fifth alarm at 2:35 a.m. Mutual aid had been called for crew relief at the evacuated stations; with the fifth alarm mutual aid companies were directed to the incident scene. The fifth alarm response was two engines, one truck company and a battalion chief.

By the time these mutual aid companies arrived, a clear perimeter had been established; numerous handlines were being used to control wide spread spot fires. Foam was being used by fireboat pumps on the north side to control flammables exposed by the intense fire as well as for application to the yachts.

Burning chemicals, smoldering fiberglass and other debris was being washed into the New River. The Coast Guard set up a floating boom to contain the runoff. Barrels containing flammable liquids used in the construction of the boats ruptured as the fire progressed, causing explosions and more burning.

A sixth alarm was called at 3:30 a.m. with two additional mutual aid engine companies called to the scene. The fire was declared out at 9:30 a.m. Mutual aid was received from four area departments, in addition to the Coast Guard and Port Everglades. Over 100 firefighters fought the fire.

Several pumpers and crews remained on site throughout the day to fight spot fires and assist in overhaul. Marina management hired front loaders to clear the debris; a team and equipment from Fort Lauderdale's Public Works Division was called in to assist.

AFTER THE FIRE

There were no Injuries or fatalities; several firefighters were treated for heat exhaustion and released.

One yacht in Building B incurred heavy smoke and water damage; two yachts in Building A were destroyed. Two aluminum hulls stored to the south of Building A were damaged. Building A and Building C were destroyed as well as the dust collector. Spot fires caused by flying brands caused minimal damage, burning the contents of several dumpsters and causing damage to some outbuildings located around the property and neighboring properties.

THE INVESTIGATION

Investigation of the fire began before the fire was declared under control. Police and fire department investigators, working in conjunction with representatives from ATF, collected numerous witness statements from marina occupants and neighbors while firefighters worked to extinguish the fire.

The investigation was organized as a joint effort between the Fort Lauderdale Fire Marshal's office, the Fort Lauderdale Police Department, the Florida State Fire Marshal's office, the United States Coast Guard, and the Bureau of Alcohol, Tobacco and Firearms. Individuals from these agencies were organized into four teams, assigned to interior investigation, exterior investigation, interviewing and evidence control. The evidence team was also responsible for photographing the scene. The interior team concentrated on the excavation and documentation of Buildings A, B and C. The exterior team documented the numerous peripheral and spot fires that occurred throughout the facility. A K-9 team from the Indian River County Department of Emergency Services was utilized to check for accelerants.

A fire watch was posted for the first 24 hours after the incident, as fire operations continued into the afternoon. The teams reassembled at 8:00 a.m. on September 6 and began the site investigation. Initially, the number of spot fires located around the site created speculation that the fire had been intentionally set. The interior team located an electrical circuit box on the west end of the mezzanine level of Building A; this was in the general location where early witness statements placed the initial flames. The box was found damaged by an electrical short and was determined to be a possible source of ignition.

The teams worked until September 9, when a final meeting was held and the team attributed the cause of the fire to an electric short in the west central end of Building A. The intensity of the fire was considered increased because of poor housekeeping practices. The multiple spot fires were determined to be the result of flying embers from the fully developed fire and subsequent radiant heat.

LESSONS LEARNED

1. **It is critical to provide a sufficient number of staff officers to allow full implementation of the Incident Command System.**

    Although it is sometimes necessary to double-up tasks on one staff officer early in an incident, such as sector officer and safety officer, it is imperative that assignments be split when additional staff becomes available. Because the safety officer must be free to survey the fire scene so as to relay information back to the incident commander, it is very difficult for this officer to also function as a sector officer.

2. **Establishing a perimeter can minimize the spread of fire and help direct resources as they arrive on the scene.**

    The facility was spread over such a large area and presented such varied risks that establishing a perimeter became a priority. Fire units were deployed to make an initial attack on buildings and exposures on the east side of the perimeter while assuring that incoming units were supplied and positioned to support the perimeter. The perimeter was in constant risk of being overrun by radiant heat and large firebrands carried aloft by the firestorm.

3. **Dispatch procedures should address the availability of specialized equipment.**

    Fort Lauderdale is a city with miles of waterways and various risks on the shores of these waterways. There was a delay in getting the fireboat on site because the crew was initially dispatched as a ladder company and had to return to quarters to respond with the boat. Specialized equipment available from mutual aid departments, such as fireboats, should also be identified in dispatch procedures and alerted early on.

4. **Pre-incident planning should incorporate information gathered through regularly scheduled fire prevention inspections.**

    The location of the second entrance to the complex and the recent construction of the intended fire wall between Buildings A and Buildings B and C were factors not immediately known to the incident commander. Ongoing fire inspections may have identified the potential problem of scaffolding surrounding yachts under construction, and information regarding how this scaffolding might perform under fire conditions could have been obtained prior to an actual incident.

5. **Water supply evaluation is an important component of pre-incident planning.**

    The failure of aged supply mains and decreased flow rates are a potential critical risk. Review of water supply capabilities should be ongoing. Procedures should address drafting operations, as waterways are an obvious asset and man made obstructions can cause problems regarding access.

6. **Delayed reporting can greatly increase the potential for property damage.**

    With no fire detection or alarm systems installed in the building of origin, the fire progressed to full building involvement before the arrival of the first engine. The first calls to 9-1-1 described a substantial fire underway; the roof of Building A had already collapsed when Engine 3 arrived. This delay between ignition and fire department arrival may have been significantly shortened if a fire detection or alarm system had been installed and operative.

7. **Coordinated efforts between investigative authorities support an efficient and prompt investigation.**

    Representatives from five separate local, State and Federal authorities worked together to conduct this investigation. Despite the split responsibility of investigation and enforcement, all team members worked together to gather information from witnesses and to conduct the investigation. A prompt investigation serves to reduce speculation from the public and permits the department to assess post fire conditions in an informed manner.

APPENDIX A

Site Plan

Area of Origin

City Water Supply

Appendix A (continued)

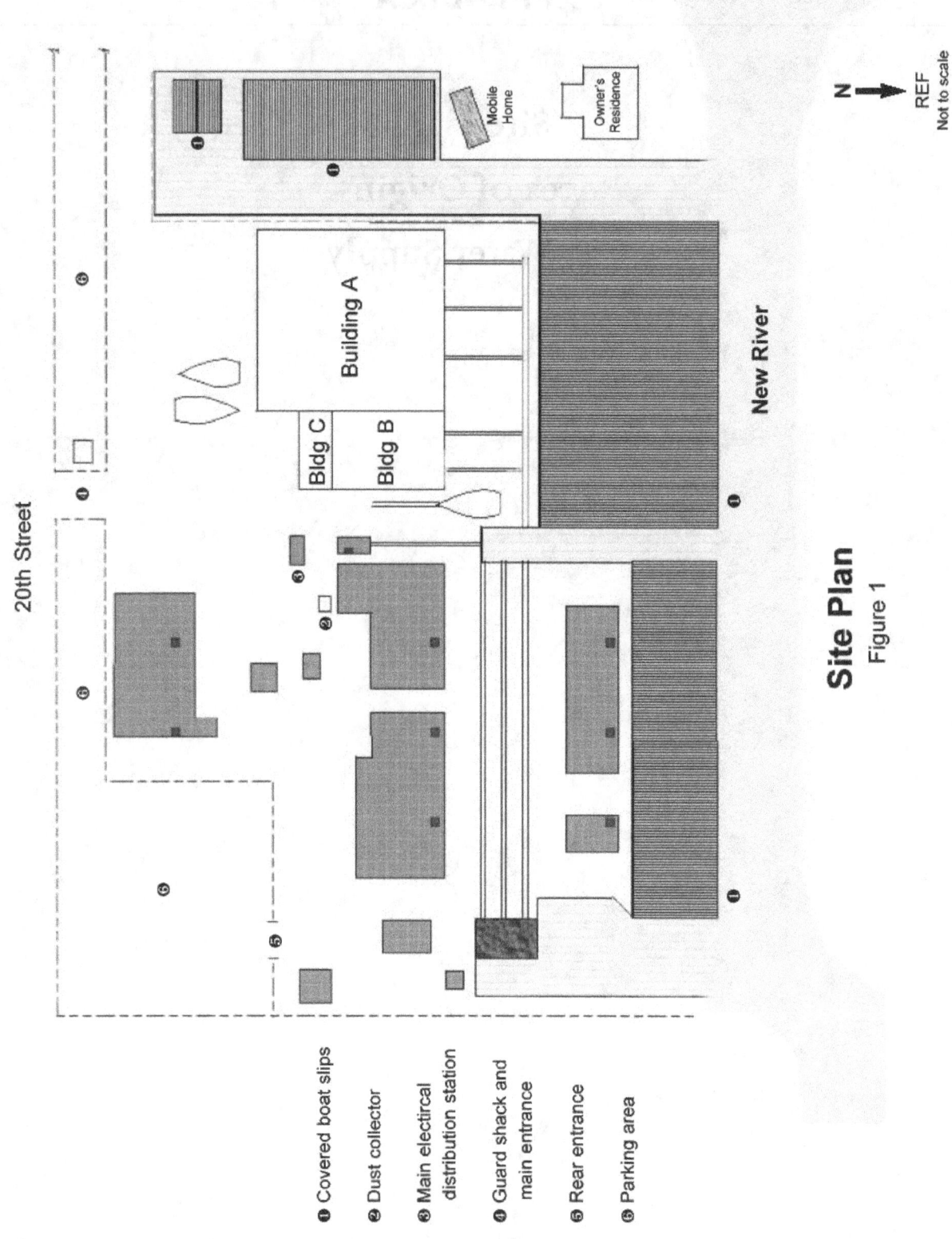

Site Plan
Figure 1

❶ Covered boat slips

❷ Dust collector

❸ Main electrical
distribution station

❹ Guard shack and
main entrance

❺ Rear entrance

❻ Parking area

Appendix A (continued)

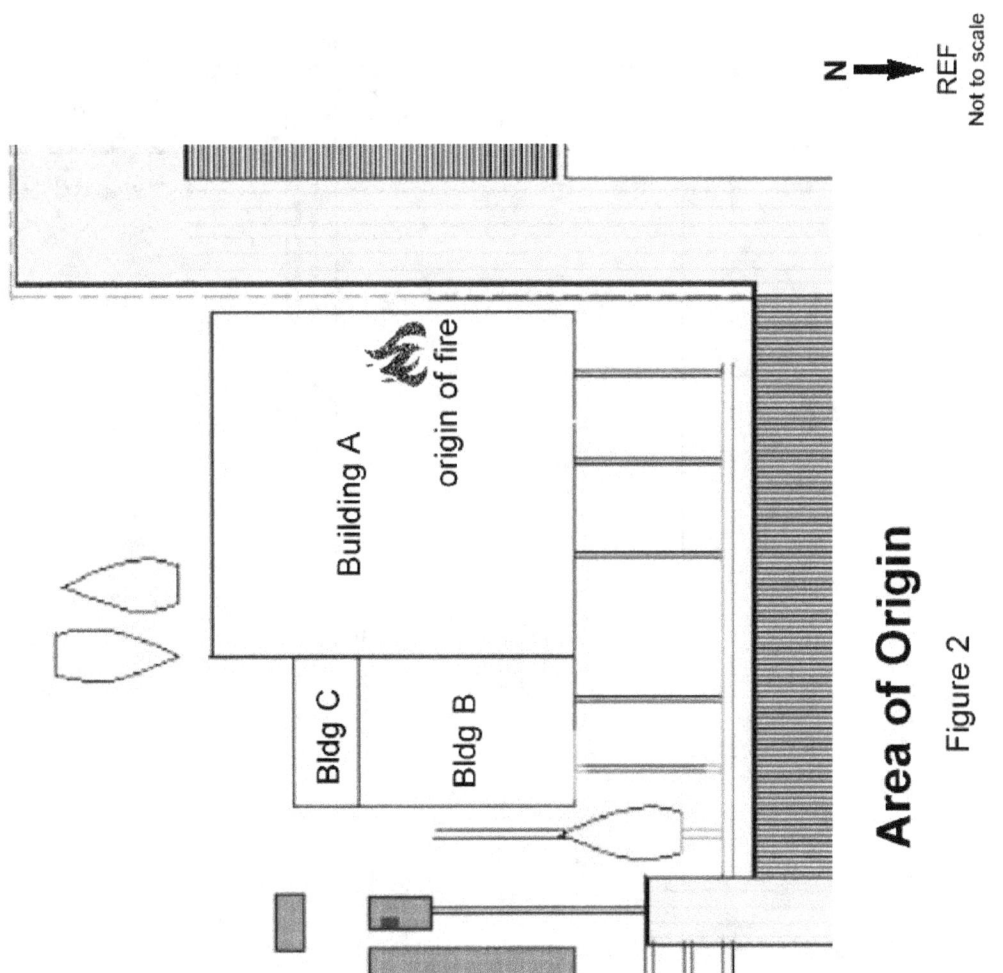

Area of Origin

Figure 2

Appendix A (continued)

Broward Marine Site

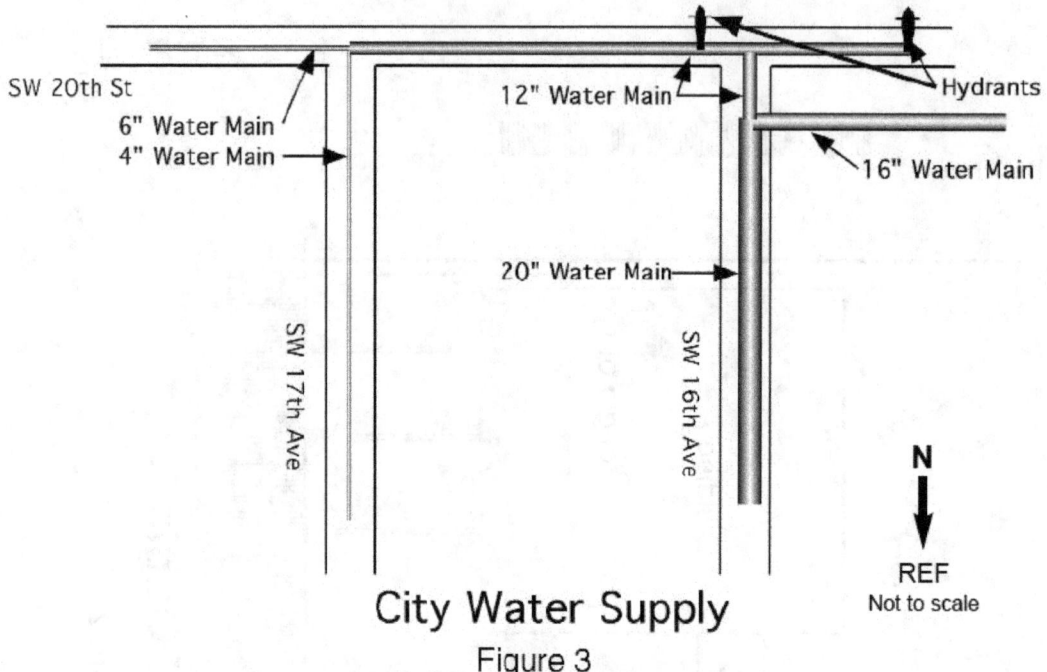

City Water Supply
Figure 3

APPENDIX B

Photographs

Photographs were obtained from the Fort Lauderdale Fire Rescue Department.

Appendix B (continued)

1. View of yacht under construction in Building B. Scaffolding constructed around the yacht hindered the firefighting efforts.

Appendix B (continued)

2. View looking east from Manufacturing Building (Building A) towards the destroyed Office Building (Building C).

Appendix B (continued)

3. Aerial view of fire scene; the two boat hulls located at the top of the photograph are situated south of Building A.

Appendix B (continued)

4. Aerial view of the south corner of the collapsed Manufacturing Building (Building A).

Appendix B (continued)

5. Aerial view of the fire scene, with the canal and neighboring covered boat slips (to the west) at top of photograph.

APPENDIX C

Breakdown of the Six Alarms

First Alarm: 12:52 a.m.
Engine Company 2
Engine Company 3
Engine Company 8
Quint Company 2
Battalion Chief 102
Battalion Chief 402

Second Alarm: 12:56 a.m.
Engine Company 29
Engine Company 49
Ladder Company 35
Battalion Chief 302
Support 13
Division Chief 2

Third Alarm: 12:59 a.m.
Engine Company 35
Engine Company 47
Ladder Company 49 (Returned to Quarters for fireboat)
Fireboat 49
Fireboat 6 (Port Everglades - Mutual Aid)
Coast Guard Fireboat (Fort Lauderdale Station - Mutual Aid)

Fourth Alarm: 01:53 a.m.
Engine Company 46
Engine Company 13

Fifth Alarm: 02:35 a.m--Mutual Aid Companies
Engine Company 5
Engine Company 32
Ladder Company 1
Battalion Chief 5

Sixth Alarm: 03:30 a.m--Mutual Aid Companies
Engine Company 30
Engine Company 38